WHEN Einstein referred to the early work of Nie
musicality in the sphere of thought," his appreciati
and a man keenly attuned to the deep creative impu
Throughout the history of science, it has been this same impulse that has led to the most significant and spectacular scientific discoveries.

Some of these discoverers were theoreticians like James Clerk Maxwell, working with nothing but "a pencil and his good Scottish brains," as the writer Eva Fenyo put it. Others were the experimentalists and practitioners who put an ever-widening human knowledge of nature's laws into practical use, and, like Ford and Edison, begat new chapters of social history that are still being written.

There are few places where the story of Western scientific development can be traced more panoramically than in the collections of the Library of Congress. From the earliest editions of Euclid and Archimedes to papers on grand unified theory, the Library's collections encompass the rich heritage of inquiry and discovery that rightly belongs to all humanity. The sampling of discoverers portrayed herein is taken from the Library's Prints and Photographs Division.

INVENTORS & SCIENTISTS

Charles Babbage (1792–1871), English mathematician and builder of calculating machines whose design anticipated many features of modern electronic computers.

Pomegranate • Box 808022 • Petaluma, CA 94975

Prints and Photographs Division, Library of Congress

INVENTORS & SCIENTISTS

Alexander Graham Bell (1847–1922), Scottish-American scientist whose work in the mechanics of speech and sound led to the invention of the telephone in 1876. One of the great inventive geniuses of the industrial age, Bell contributed throughout his life to a wide range of fields from sheep-breeding to aerodynamics.

Prints and Photographs Division, Library of Congress

Pomegranate • Box 808022 • Petaluma, CA 94975

INVENTORS & SCIENTISTS

Niels Bohr (1885–1962), Danish physicist whose application of quantum statistical methods to atomic structure yielded a new atomic model in 1913. Bohr's sheer personal and intellectual force had a great impact on the development of 20th century physics and its leading innovators.

Pomegranate • Box 808022 • Petaluma, CA 94975

Prints and Photographs Division, Library of Congress

INVENTORS & SCIENTISTS

Luther Burbank (1849–1926), American plant breeder who developed over 800 new varieties of fruits, flowers, vegetables, grains, grasses and forage plants, many of which are still economically important.

Pomegranate • Box 808022 • Petaluma, CA 94975

INVENTORS & SCIENTISTS

Marie Curie (1867–1934), Polish-French chemist whose isolation of polonium and radium (with her husband Pierre) marked the beginning of a new era in the study of atomic structure. Awarded Nobel Prizes in Physics (1903) and Chemistry (1911). Shown here touring an American chemical plant in 1921.

Pomegranate • Box 808022 • Petaluma, CA 94975

Prints and Photographs Division, Library of Congress

INVENTORS & SCIENTISTS

Charles Robert Darwin (1809–82), British naturalist who revolutionized the science of biology by his demonstration of evolution by natural selection, an idea that spread into all branches of human thought and continues to upset established patterns of thinking.

Pomegranate • Box 808022 • Petaluma, CA 94975

Prints and Photographs Division, Library of Congress

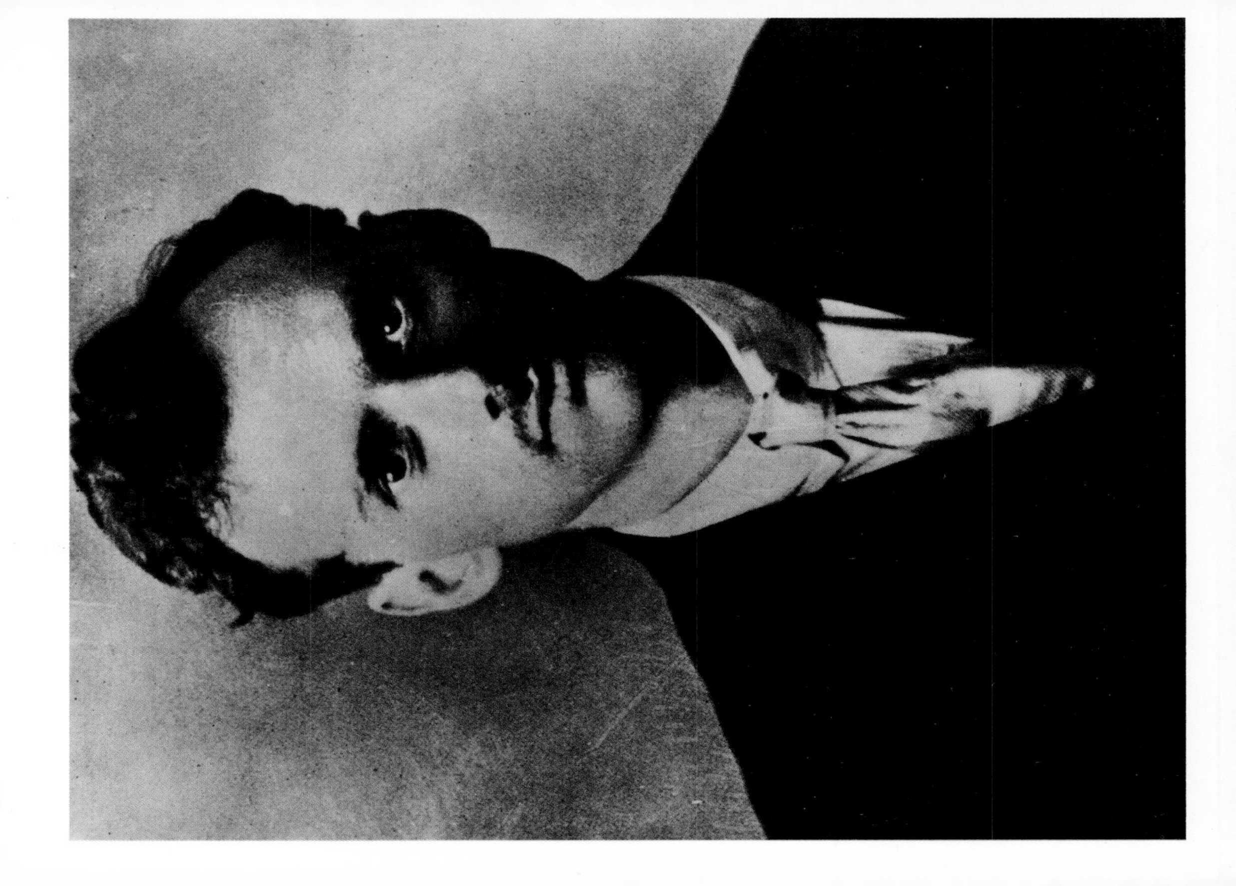

INVENTORS & SCIENTISTS

Paul Adrien Maurice Dirac (1902–), British theoretical physicist who made numerous contributions to the quantum mechanical description of atomic phenomena. In his theory of pair production Dirac predicted the first of the antiparticles, the positron, whose existence was subsequently confirmed by experiment.

Pomegranate • Box 808022 • Petaluma, CA 94975

INVENTORS & SCIENTISTS

George Eastman (1854–1932), American industrialist best known for his innovations that brought photography within the reach of the masses. Eastman was also a noted philanthropist, giving some $75 million to the University of Rochester, MIT, and the Hampton and Tuskegee Institutes.

Pomegranate • Box 808022 • Petaluma, CA 94975

Prints and Photographs Division, Library of Congress

INVENTORS & SCIENTISTS

Thomas Alva Edison (1847–1931), American inventor and pioneer industrialist. Rough-hewn and self-educated, Edison rose to become a folk hero in his day and one of the most prolific inventors of all time. Three of his inventions—the phonograph, the practical incandescent light, and the movie camera—spawned giant industries that were to change the life and leisure of the world.

Pomegranate • Box 808022 • Petaluma, CA 94975

Prints and Photographs Division, Library of Congress

INVENTORS & SCIENTISTS

Albert Einstein (1879–1955), German-American scientist and one of the greatest theoretical physicists of all time. Best known for the special and general relativity theories, Einstein was also a pioneer of quantum theory and made important contributions to the kinetic theory of matter and the theory of specific heats. He was also a renowned humanitarian and an accomplished amateur violinist.

Pomegranate • Box 808022 • Petaluma, CA 94975

Prints and Photographs Division, Library of Congress

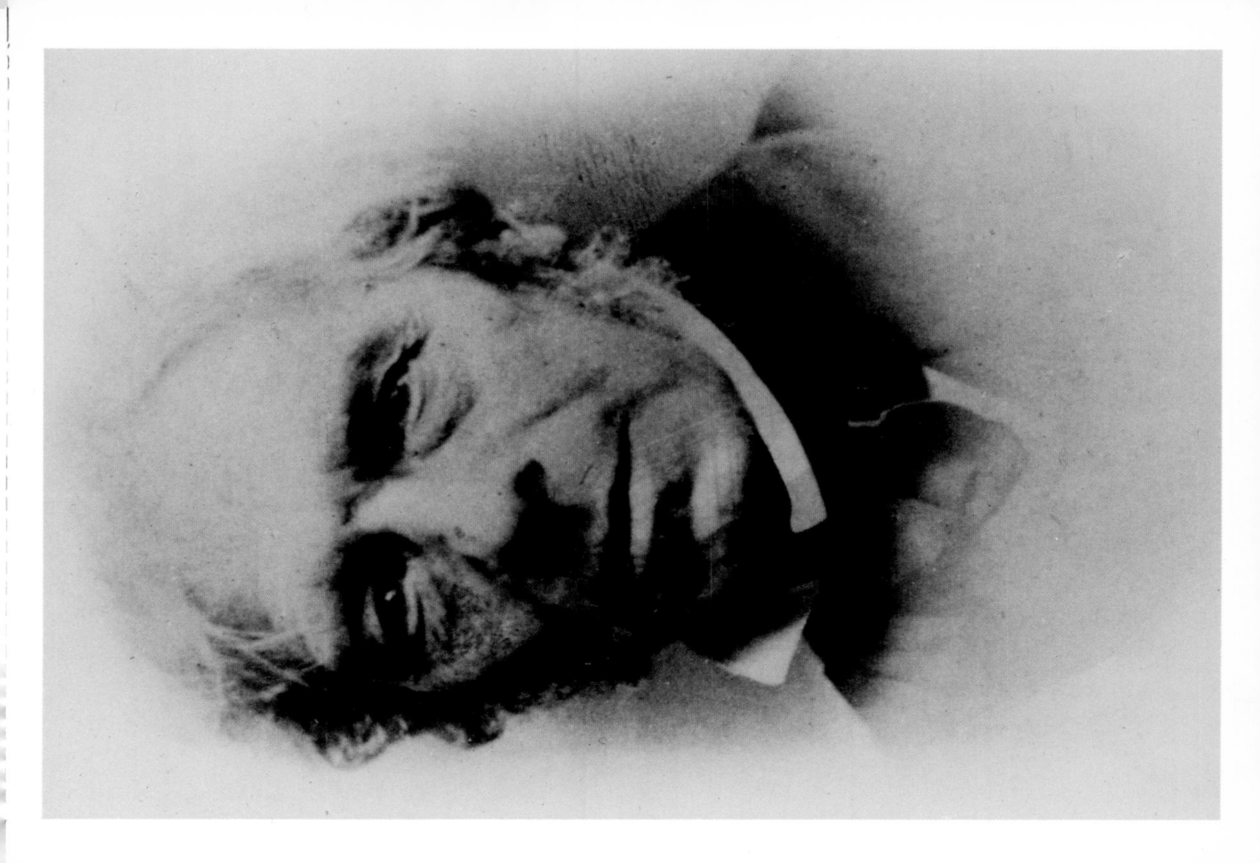

INVENTORS & SCIENTISTS

Michael Faraday (1791–1867), English physicist and chemist who discovered electromagnetic induction and founded the science of electrochemistry. Widely regarded as one of the greatest experimental scientists of all time, Faraday was also a brilliant theorist and was the principal architect of the classical field theory later developed by Maxwell and Einstein.

Pomegranate • Box 808022 • Petaluma, CA 94975

Prints and Photographs Division, Library of Congress

INVENTORS & SCIENTISTS

Henry Ford (1863–1947), autocratic American manufacturer whose development of the first popular low-priced automobile and the technology of the moving assembly line were to forever change modern industry and society.

Pomegranate • Box 808022 • Petaluma, CA 94975

Prints and Photographs Division, Library of Congress

INVENTORS & SCIENTISTS

Galileo Galilei (1564–1642), Italian astronomer and mathematician who helped establish the modern scientific method through his use of observation and experimentation. His work in mathematics, physics and astronomy made him a leading figure of the early scientific revolution and a repeated target of the clerical establishment.

Pomegranate • Box 808022 • Petaluma, CA 94975

Prints and Photographs Division, Library of Congress

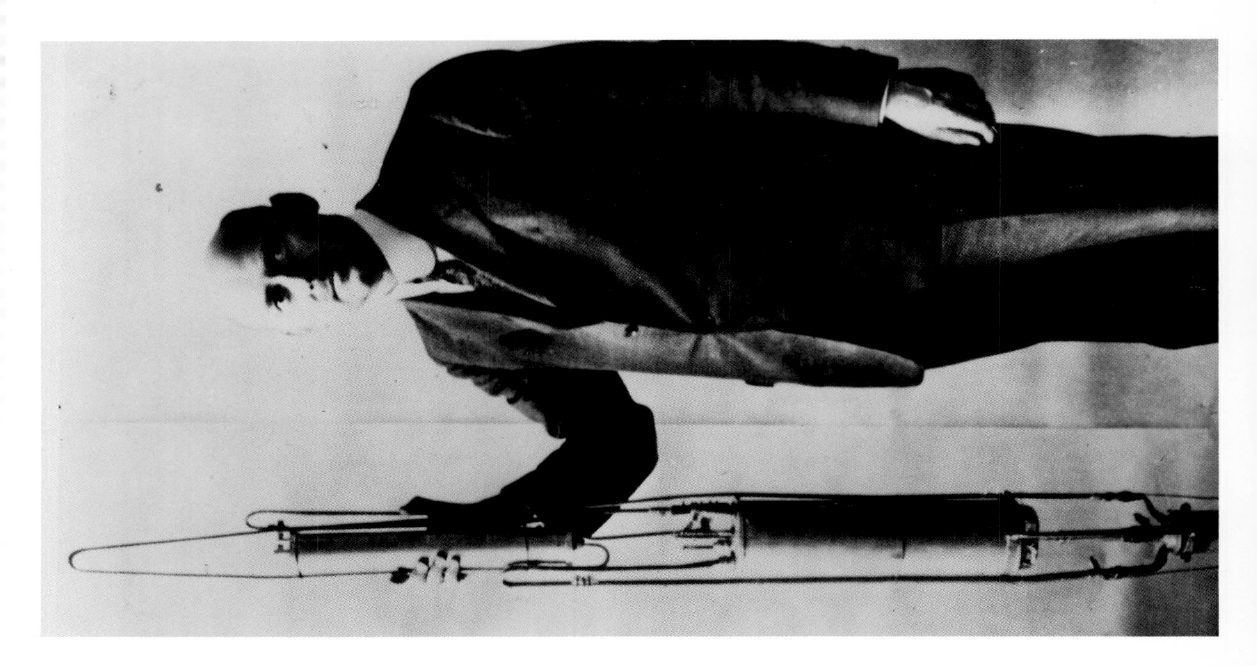

INVENTORS & SCIENTISTS

Robert H. Goddard (1882–1945), American physicist, pioneer of rocketry and tireless explorer of its theoretical and practical problems. His first successful liquid fuel rocket flight on March 16, 1926, marked the beginning of the age of space exploration.

Pomegranate • Box 808022 • Petaluma, CA 94975

Prints and Photographs Division, Library of Congress

INVENTORS & SCIENTISTS

Johann Gutenberg (1394–1468), German craftsman and inventor of printing with moveable type. The so-called Gutenberg Bible (c. 1458), of which several copies are extant, remains an unparalleled example of beautiful bookmaking and one of technology's most amazing cases of near perfection on a first try.

Pomegranate • Box 808022 • Petaluma, CA 94975

Prints and Photographs Division, Library of Congress

INVENTORS & SCIENTISTS

Sir William Herschel (1738–1822), English astronomer, discoverer of the planet Uranus and early investigator of double-star systems. His sister Caroline shared in all his astronomical work and is known as the first important woman astronomer.

Pomegranate • Box 808022 • Petaluma, CA 94975

Prints and Photographs Division, Library of Congress

INVENTORS & SCIENTISTS

Heinrich Rudolf Hertz (1857–94), German physicist who was the first to confirm experimentally the existence of electromagnetic radiation predicted by Maxwell in 1864. His work provided the basis for modern radio, wireless and television.

Pomegranate • Box 808022 • Petaluma, CA 94975

Prints and Photographs Division, Library of Congress

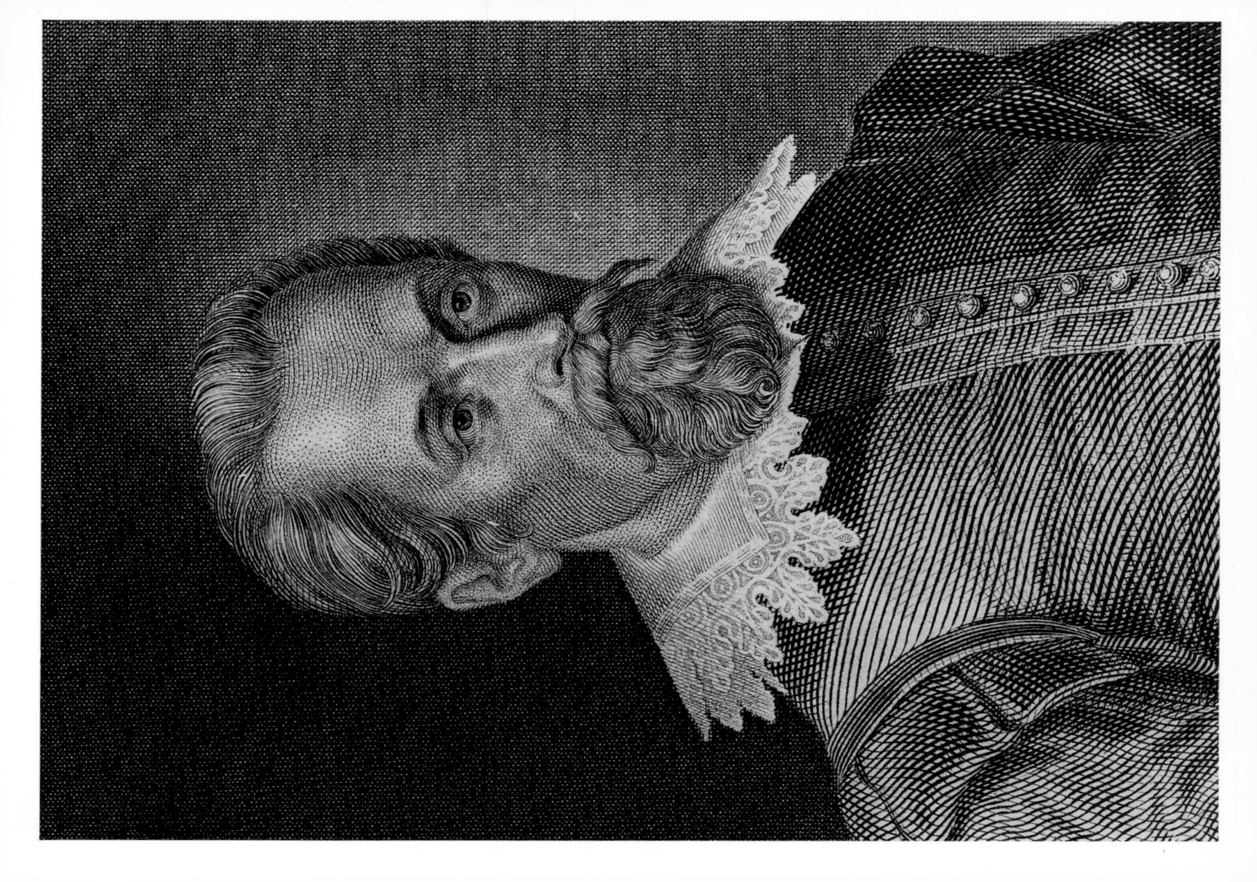

INVENTORS & SCIENTISTS

Johannes Kepler (1571–1630), German astronomer and physicist whose mathematical proofs of the Copernican system marked the beginning of modern astronomy and laid the groundwork for Newton's laws of universal gravitation.

Pomegranate • Box 808022 • Petaluma, CA 94975

Prints and Photographs Division, Library of Congress

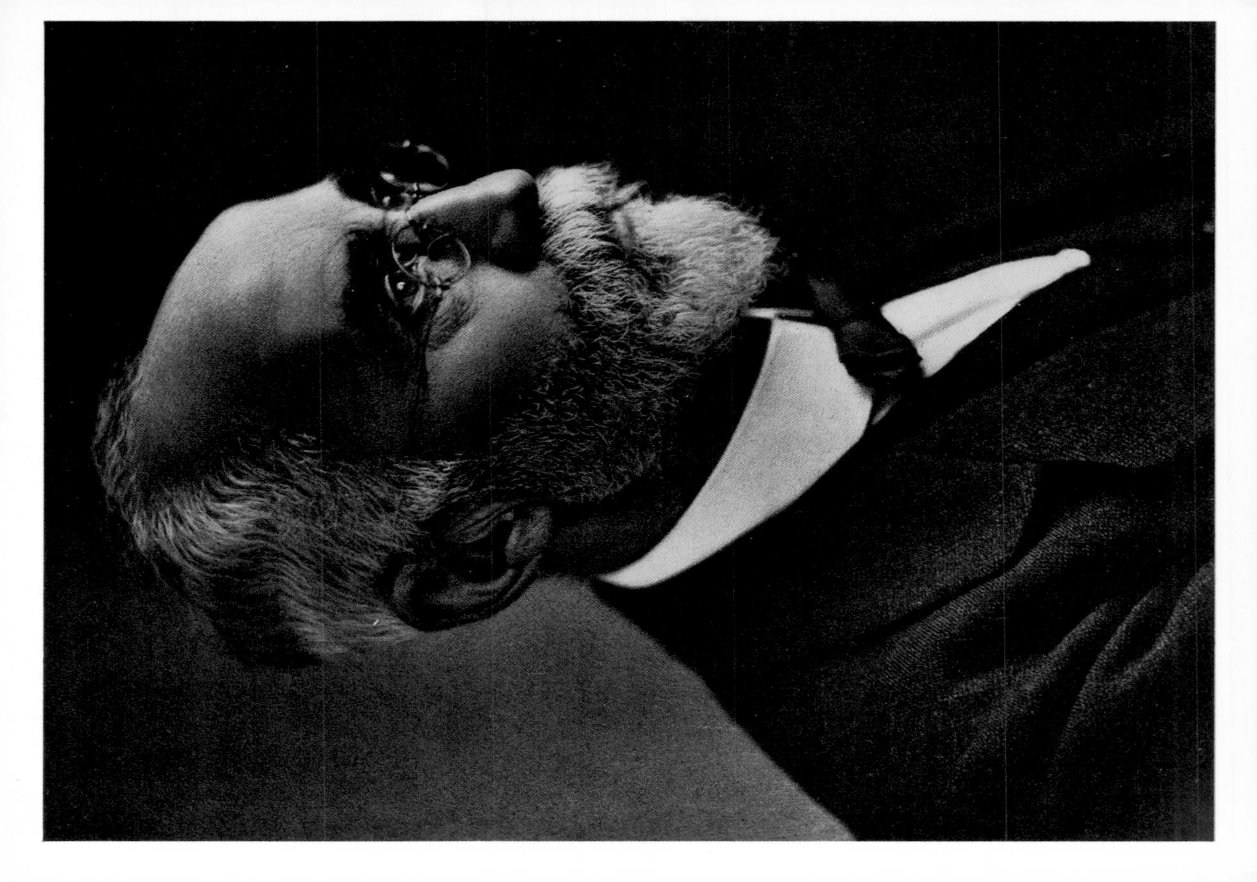

INVENTORS & SCIENTISTS

Hendrik A. Lorentz (1853–1928), Dutch physicist whose work on electromagnetic theory paved the way for the development of relativity and quantum mechanics.

Pomegranate • Box 808022 • Petaluma, CA 94975

Prints and Photographs Division, Library of Congress

INVENTORS & SCIENTISTS

Guglielmo Marconi (1875–1937), Italian physicist and inventor who first devised the practical means by which electromagnetic waves could be employed in telegraphic communication. He was awarded the Nobel Prize in Physics in 1909.

Pomegranate • Box 808022 • Petaluma, CA 94975

Prints and Photographs Division, Library of Congress

INVENTORS & SCIENTISTS

James Clerk Maxwell (1831–79), Scottish physicist whose mathematical descriptions of interacting magnetic and electric fields in the emission of radiant energy confirmed the existence of electromagnetic waves. Einstein was to use Maxwell's equations as a basis for a central postulate of special relativity—the speed-of-light limitation.

Pomegranate • Box 808022 • Petaluma, CA 94975

INVENTORS & SCIENTISTS

Lise Meitner (1878–1968), Austrian-Swedish physicist whose nuclear research and discovery of new elements (primarily with Otto Hahn) paved the way for atomic fission, a term she coined in 1939 upon correctly interpreting the results of neutron bombardment experiments of Hahn and Fritz Strassman.

Pomegranate • Box 808022 • Petaluma, CA 94975

Prints and Photographs Division, Library of Congress

INVENTORS & SCIENTISTS

Isaac Newton (1642–1727), English alchemist, theologian, recluse, and dogged experimenter, whose 1687 *Philosophiae Naturalis Principia Mathematica* still stands as one of the most remarkable intellectual syntheses of all time.

Pomegranate • Box 808022 • Petaluma, CA 94975

Prints and Photographs Division, Library of Congress

INVENTORS & SCIENTISTS

Louis Pasteur (1822–95), French biologist and chemist, best known for his work in bacteriology, a field he almost single-handedly founded. Pasteur's contributions to the control and treatment of disease (many of which were achieved after a devastating stroke at age 45) were myriad and had a lasting impact on many industries and methods.

Pomegranate • Box 808022 • Petaluma, CA 94975

Prints and Photographs Division, Library of Congress

INVENTORS & SCIENTISTS

Linus Pauling (1901–), American chemist and peace activist whose 1939 *The Nature of the Chemical Bond and the Structure of Molecules and Crystals* is widely regarded as one of the most influential scientific treatises of the century. Pauling helped revolutionize and found several scientific disciplines and is the only person to have won two unshared Nobel Prizes (one of Mme. Curie's was shared), in Chemistry and Peace.

Prints and Photographs Division, Library of Congress

Pomegranate • Box 808022 • Petaluma, CA 94975

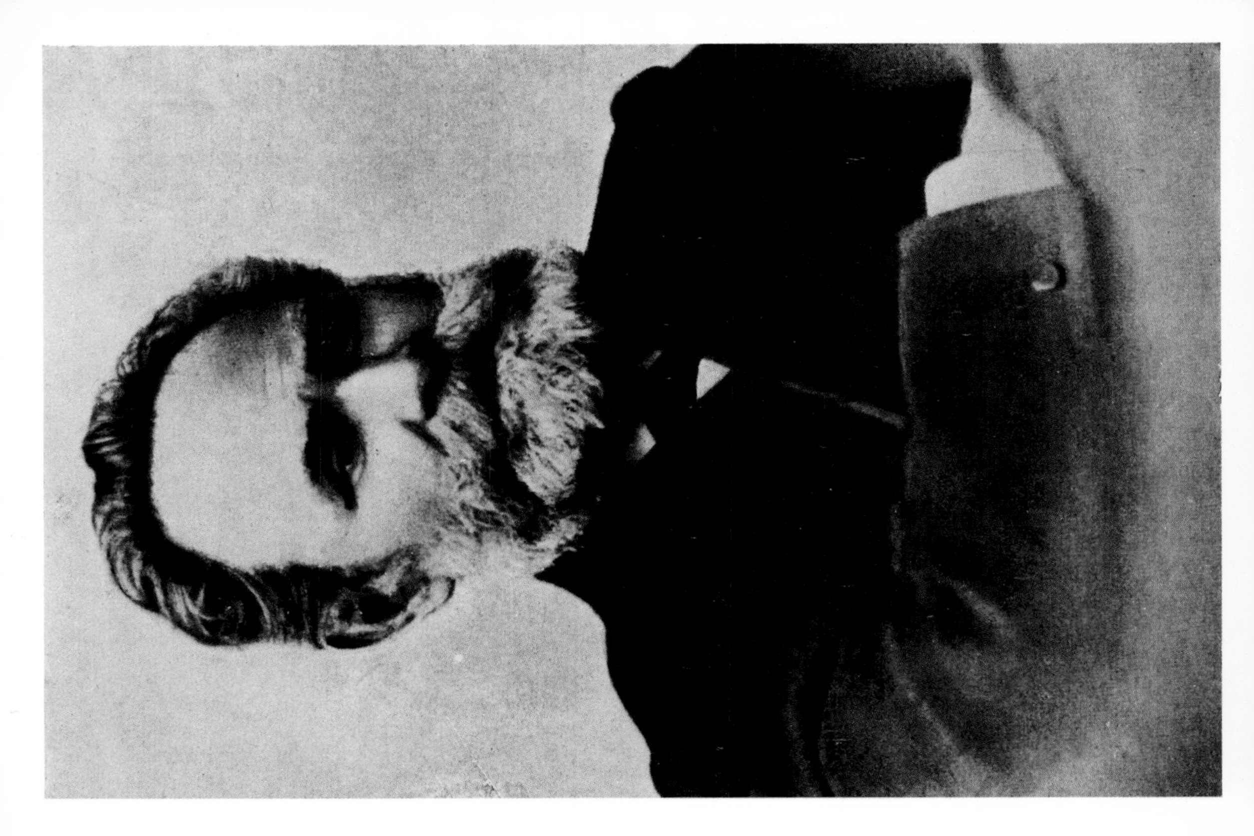

INVENTORS & SCIENTISTS

Ivan Petrovich Pavlov (1849–1936), Russian physiologist best known for developing the theory of conditioned reflexes. Pavlov also conducted important experiments on the relation between behavior and the nervous system.

Pomegranate • Box 808022 • Petaluma, CA 94975

Prints and Photographs Division, Library of Congress

INVENTORS & SCIENTISTS

Max Planck (1858–1947), German physicist generally regarded as the founder of quantum theory. His famous constant *h*, employed as "an act of desperation" in describing energy-to-frequency relations, has proven a cornerstone of quantum physics and one of the most universal constants in all nature.

Pomegranate • Box 808022 • Petaluma, CA 94975

INVENTORS & SCIENTISTS

Ernest Rutherford (1871–1937), British physicist whose pioneering work in atomic structure and the nature of radioactivity led to his landmark nuclear model of the atom in 1904. Equally brilliant as a theorist and experimenter, Rutherford was also an early influence on the young Niels Bohr.

Pomegranate • Box 808022 • Petaluma, CA 94975

Prints and Photographs Division, Library of Congress

INVENTORS & SCIENTISTS

Orville Wright (1871–1948) and Wilbur Wright (1867–1912), Americans from whose bicycle business in Dayton, Ohio, grew the seeds of modern aviation. Initially stimulated by a toy helicopter, the brothers' experiments with gliders in 1899 led to their famous first powered flight in 1903.

Pomegranate • Box 808022 • Petaluma, CA 94975

Prints and Photographs Division, Library of Congress